Guilherme Cardoso

# Fishing's contribution to social and economic development

Guilherme Cardoso

# Fishing's contribution to social and economic development

## Fishing and its contribution to social and economic development in the city of Itajaí

ScienciaScripts

**Imprint**



Cover image: www.ingimage.com

This book is a translation from the original published under ISBN 978-613-9-68960-6.

Publisher:
Sciencia Scripts
is a trademark of
Dodo Books Indian Ocean Ltd. and OmniScriptum S.R.L publishing group

120 High Road, East Finchley, London, N2 9ED, United Kingdom
Str. Armeneasca 28/1, office 1, Chisinau MD-2012, Republic of Moldova, Europe
Printed at: see last page
**ISBN: 978-620-8-24192-6**

## SUMMARY

## ACKNOWLEDGMENTS

To the PROESDE Course Coordinator Prof.ª Samara, for the effort and enthusiasm with which she runs this course.

To my advisors Professors Ayrton Santos and Josiane da Luz for sharing their time and their time and knowledge, and for their attention.

To all the professors who taught classes, bringing us knowledge to help us develop this project.
this project.

To Mr. Nicélio Assi Veloso for sharing all his professional experience acquired over the years as a fisherman.

To Mr. Agostinho Peruzo for all his collaboration and attention in passing on to us some of his knowledge on the subject.

To the government of the State of Santa Catarina for encouraging social projects and for the opportunity they have given us.

To everyone who contributed in any way to the project's completion.

UNIVALI for giving us the opportunity to take part in the PROESDE project.

To my colleagues for their companionship and friendship.

To all of you, thank you very much.

I would like a fairer, less corrupt society, with less hypocrisy, more dignity, more love for others, less prejudice, less resentment and, above all, more peace in the soul.

(Albert Einstein)

**TECHNICAL TEAM**

a) Academics

GUILHERME A FABRICIO CARDOSO

LARISSA NUNES

LUISA BARBIER SANTOS

ROZANGELA APARECIDA DE MELO GONÇALVES

b) Research area

Valuing and preserving fishing in the city of Itajai

c) Content supervisor Prof.ª Msc Josiane da Luz Prof.ª Msc Ayrton Santos

d) Course leader

Prof. MSc Samara Aparecida da Silva Garcia

## SUMMARY

The aim of the project was to revive Itajai's historical and cultural heritage, which is fishing. Due to its excellent coastal location, the municipality of Itajai has an important infrastructure for the growth and development of the fishing sector. The city of Itajai is an important fishing center which, together with the other coastal cities of Santa Catarina, accounts for 90% of all industrial fishing production in Santa Catarina. As fishing is one of the oldest and most important professions for the livelihood of many families, it is essential that the activity is valued and stimulated in order to reach future generations. Some of the bodies responsible for the fishing sector have been losing ground and credibility. The Ministry of Fisheries itself has been abolished and is now under the responsibility of the Ministry of Agriculture and Fisheries, so it is up to society in general to cultivate and value this profession. With fishing interfering in various areas such as the economy, culture, gastronomy and tourism, this project proposes training for teachers and educational actions aimed at raising awareness among elementary school pupils of the importance and need to know and preserve the history and contributions of fishing to the municipality, through the introduction of fishing into everyday school life by the teachers themselves.

**Keywords:** Fishing culture. Educational actions. Student awareness.

# 1 INTRODUCTION

Over the years, the subject of fishing has ceased to be treated with due importance, thus losing a good part of the rich history of the municipality of Itajai, which was once recognized as the Brazilian capital of fishing. Today, the recognition and appreciation of this ancient profession, which has guaranteed and contributed to the growth and development of the municipality, has ceased.

To realize the importance of fishing today is to value the culture and history of Itajai, recovering its essence and giving the current municipalities a chance to learn a little more about the contribution of the fishing sector not only to the locality, but to the entire state of Santa Catarina and also to Brazil.

The historical facts that point to all this development require greater attention in order to raise awareness and stimulate the sector so that present and future generations can participate in and benefit from this heritage in a direct and indirect way, since it drives various businesses and different branches of the locality.

Currently, the municipality of Itajai has significant figures, such as "around 50 processing companies that produce more than a million cans of sardines and tuna a day. There are 250 shipowners and a fleet of 500 boats that account for 20% of Brazil's fish production, estimated at 500,000 tons a year. Itajai's fishing production is concentrated on tuna, sardines and shrimp. Around 15,000 people work directly and indirectly in the city's fishing industry. "[1]

One event that had a major negative impact on the fishing movement was the extinction of the Ministry of Fisheries, which brought uncertainty and insecurity to the entire sector, leaving gaps that are unlikely to be filled. All this lack of support and appreciation has had a significant impact, causing even more problems and preventing the strengthening and growth of fishing in recent years.

Given the importance of the subject, the project's main objective is to expand the culture of fishing, valuing and raising awareness in the community through children's education, introducing this subject, which is so important to the history of the city of Itajai, back into the daily lives of students in the classroom, so that this tradition can be strengthened and

---

1 Itajai, between the river and the sea. Available at: <http://itajai.sc.gov.br/c/a-cidade#.WXzD1oTyvIU>. Accessed on July 29, 2017.

developed even further. To do this, it is essential to train education professionals who will be the actors introducing the culture and appreciation of fishing into their students' daily lives.

Public policies have become important tools for transformation and can innovate and benefit hundreds of people, through actions aimed at the community, with a view to development and social well-being. As Jani Alves da Silva Moreira and Angela Mara de Barros Lara point out[2] "recent educational policies in Brazil and other countries are marked by specific discourses on early childhood care as a priority for the sustainable development of countries".

In this way, the aim is to raise awareness among the entire community through actions aimed at elementary school, seeking to arouse interest and highlight the importance of fishing culture for regional growth and development.

It is also important to highlight the teachings of Fabio Pinto Gonçalves dos Reis and Delcimar de Oliveira Cunha :[3]

> It is of fundamental importance to consider that, since the Federal Constitution of 1988, the education of children from 0 to 6 years old in Early Childhood Education (nurseries and pre-schools) has been established as a right. The Statute of the Child and Adolescent (1990) and the Law of Guidelines and Bases of National Education (LDBEN, 1996) corroborate this legislation by integrating it into the educational system, as the first stage of Basic Education, also comprising Primary Education - which is compulsory - and Secondary Education.

Understanding the important role that education plays for society in general, we are seeking to consolidate our region's fishing culture in primary education as a way of creating and strengthening an educational model that contributes to regional development.

---

2 MOREIRA, Jani Alves da Silva, and LARA, Angela Mara de Barros. **Public policies for early childhood education in Brazil (1990-2001)** [online]. Maringa: Eduem, 2012. pp. 89-90. ISBN 978-85-7628585-4. Available from SciELO Books.

3 PIMENTA, Carlos Alberto Maximo., and ALVES, Cecilia Pescatore., orgs. **Politicas pùblicas & desenvolvimento regional** [online]. Campina Grande: EDUEPB, 2010. p. 106. ISBN 978-85-7879-0165. Available from SciELO Books .

## 2. CONTEXTUALIZATION OF THE MUNICIPALITY

Itajai has a surface area of 288,286 km2 and a population density of 636.11 km2. Itajai is located on the north-central coast of Santa Catarina, at the mouth of the Itajai-Açù river, in the tourist region known as Costa Verde & Mar. It borders Balneàrio Camboriù, Camboriù, Brusque, Gaspar, Ilhota and Navegantes. With an estimated population of 208,958 according to the 2016 census.

Itajai's Human Development Index (HDI) was 0.795 in 2010, placing the municipality in the High Human Development range (HDI between 0.700 and 0.799). The dimension that contributes most to the municipality's MHDI is Longevity, with an index of 0.884, followed by Income, with an index of 0.778, and Education, with an index of 0.730.

Itajai's emancipation began in 1860, when the Villa do Santissimo Sacramento do Itajai was established. It was at this time that the town hall was set up and the pillory was erected - a symbol of justice and local power.

According to historian Oswaldo Cabral, Itajai is magnificently situated at the entrance to the Itajai valley and has a practical port, which gave rise to an urban center. This happened more definitively on June 15, 1860, with the establishment of the municipality of Itajai, dismembered from Porto Belo. At the time, the mayor was in charge of executive functions. The figure of the superintendent only appeared in the Republic. This period also saw the creation of the Comarca (1868) and the town's elevation to city status in May 1876. It was at this time that the first Catholic church was built in the village. According to Antônio da Costa Flores, the church was built by the hands of Agostinho Alves Ramos' slave, Simeâo.

Flores recalled that, when he came to live in the town, "on the land that the current perimeter of this town covers and which, as you know, reaches two kilometers from the Mother Church, on all sides, except for the river which is a very short distance away, there were about fifty houses, including miserable little ranches, which, apart from being covered in straw, consisted of a single room with walls made only of slats nailed together.

It's not surprising that important historians emphasize Itajai's privileged geography. Oswaldo Cabral has often referred to the importance of the port in shaping the city. Even before it was founded, the lands bathed by the sea and river were gateways for immigrants and it didn't take long for the port to become the main means of boosting the economy. More than a means of mobilizing the flow of the economy, the port is a place of sociability. In 1906, a year after the Minister of Transportation, Lauro Müller, approved projects that would bring improvements to the port, the XV de Novembro Beneficent Society was created by port workers, which in 1939 was transformed into the Itajai Trapiche and Warehouse Workers' Union.

Over the course of the 20th century, the town became increasingly intertwined with the port. Political movements, cultural activities, workers. The decades passed and Itajai, with its port, became increasingly international. Today, Itajai has a population of 147,000 inhabitants of German and Azorean descent .[4]

Itajai is located on the north-central coast of the state of Santa Catarina at the mouth of the Itajai-Açu River. The city is located in a strategic region of Santa Catarina; on the right bank of the state's largest river, it is part of an axis connecting it to Brazil's main highways and airports .[5]

The municipality has the 3rd largest GDP in the state, due to the most important economic sectors such as logistics, construction, industrial activities and the port complex, which is the second largest in the country in terms of container handling. Itajai is constantly growing, especially with new markets expanding, such as the oil, gas and nautical industries. With over 200,000 inhabitants, it is also a place of historical, cultural and traditional riches. Throughout the year, the city maintains its tradition of typical festivals, which offer gastronomy, culture and entertainment to locals and tourists alike.

Itajai also has beautiful natural landscapes, from the countryside to the beaches, crowned by tourist attractions such as Bico do Papagaio (Figure 1), Parque Natural da Atalaia (Figure 2), Praia do Molhe (Figure 3), among others. With so much natural beauty, there is no shortage of adventure sports for all tastes: hiking, surfing, cycle tourism, open-air flying, hiking, abseiling .[6]

Figure 1: Bico do Papagaio

4 Available at: <http://www.litoraldesantacatarina.com/itajai/caracteristicas-de-itajai.php>. Accessed on 02 Oct. 2017.
5 Available at: <http://visiteitajai.com.br/conheca_acidade.php>. Accessed on October 2, 2017.
6 Available at: <http://www.mgmoperadora.com.br/destino/2285/1/28/destinos-nacionaisitajai>. Accessed on October 2, 2017.

Source: http://turismo.sc.gov.br .[7]

It is a rocky formation, carved out of nothing, which resembles the figure of a parrot's beak. It is one of the city's postcards and is located just off Praia da Atalaia in the direction of Cabeçudas beach.

Figure 2: Atalaia Natural Park

Source: http://osoldiario.clicrbs.com.br/sc/ .[8]

A natural space with a beautiful landscape, it offers outdoor recreation for adults and children

7 Available at: <http://turismo.sc.gov.br/institucional/index.php/pt-br/component/phocagallery/51-costa-verde-e-mar-itajai/detail/359-bico-do-papagaio-itajai-rec0034#>. Accessed on September 20, 2017.
8 Available at: <http://osoldiario.clicrbs.com.br/sc/noticia/2013/01/parque-da-atalaia-em-itajai-e-opcao-para-quem-quer-fugir-do-agito-do-litoral-4006901.html>. Accessed on September 20, 2017.

and a great option for tourism, education, scientific research, free flights and an exuberant landscape.

Figure 3: Molhe Beach

Source: http://turismo.sc.gov.br .[9]

The pier beach is excellent for surfing, with kiosks by the sea, showers, parking and a beautiful view of the meeting of the Itajai-Açù River and the Atlantic Ocean. Another great attraction is the special lighting for night surfing, making the place even more attractive for visitors. In addition to its natural beauty, the municipality of Itajai also stands out for its strong economy, which has boomed in recent years, reaching the highest GDP in the state in 2012[10] . This rise represents Itajai's economic potential and plurality and shows that the fishing industry's contribution to the city continues to be very important.

9 Available at: <http://turismo.sc.gov.br/institucional/images/phocagallery/costa%20verde%20e%20mar%20-%20itajai/thumbs/phoca_thumb_l_Vista%20A%C3%A9rea%20Molhes%20E%20%20Praia%20Do%20Atalaia%20- %20Itajai5701.jpg>. Accessed on September 20, 2017.

10 Available at: <https://ndonline.com.br/joinville/noticias/calculo-do-pib-catarinense-e-revisado-pelo- ibge-e-coloca-joinville-itajai-e-florianopolis-no-topo>. Accessed on September 21, 2017.

# 3. OBJECTIVES

**3.1** General Objective

> Create a program for elementary school teachers to raise awareness of fishing culture in the municipality of Itajai/SC.

**3.2** Specific Objectives

> Present the importance and contribution of fishing to the development of the local economy;

> To identify the perception of future teachers in relation to fishing culture;

> Present teaching actions aimed at enhancing the value of fishing.

**3.3** Justification for carrying out the study

The topic "FISHING" was chosen because it is extremely relevant to Itajai's economic scenario. In order to understand the importance of fishing today, it is necessary to look back over the centuries, when fishing was initially a subsistence activity, i.e. man used this extractive activity simply to supplement his food without selling or bartering. Over the years and later with the emergence of communities, towns and cities, man then began to produce food for sale and to generate wealth.

The relevance of the issue is all the more impressive given the latest figures, which show that the municipality accounts for 59% of Santa Catarina's fishing production, making it the country's main fishing center. The state accounts for 14% of national establishments and 16% of all jobs.

The sector also involves a wide range of complementary activities. Among them, the state stands out nationally in the construction of boats and floating structures, being the third with the largest number of establishments and the fourth in terms of jobs, with a national share of approximately 12% in both respects.

Thus, the relevance of the issue for the region is evident, as it is of great importance at state and federal level, but managers need to be aware of the need to preserve this culture so that future generations can enjoy this wealth.

The current problems facing this sector go far beyond legal obstacles. In recent years, fisheries management has made little progress, the result of inertia on the part of the federal government, successive changes and, above all, the end of the "Ministry of Fisheries", announced in 2015 by then-president Dilma Rousseff, which has since been merged with the Ministry of Agriculture.

It is understood that there is a need for policies committed to this category that contributes so much to the economic growth of the city and the country. According to recent figures, fishing accounts for around 90.95% of jobs in Santa Catarina, with 3,016 workers and a large number of companies linked to fishing, which boost other sectors. However, each year the number of fishermen has been decreasing and the market weakening due to a lack of expectation in the fishing sector .[11]

The main focus of this research is the problems faced by industrial fishing in relation to the closed season, which, by means of ordinances, limits fishing activities, bringing numerous limitations and problems for fishermen. These measures were first introduced with the idea of protection, given that many of the species fished are, according to the Ministry of the Environment and Ibama, endangered.

As a result, the project will have the positive points of facilitating and implementing measures that make it possible to protect all endangered species, as well as continuing to provide growth for the sector, expanding it and consequently generating jobs and income.

As such, the project provides the opportunity to study and research the subject in a different way to how it has been dealt with so far, since the University, through us academics, will be able to implement measures based on the studies and research carried out, bringing development and countless benefits to society as a whole. From this point of view, if a sector is to be strong and remain consolidated, it needs to be valued; fishing needs to be stimulated if it is to be able to effectively develop and reach the next generations.

The sector can continue to make a more significant contribution to generating new jobs, thus contributing significantly to the regional economy and to the education of our children and teenagers. Benefits such as better qualifications for professionals and recognition from the community are some of the effects that the project will bring to all those who make a living

11 Available at: <https://ndonline.com.br/florianopolis/noticias/producao-registrada-entre-2010-e-2013-mantem-santa-catarina-na-lideranca-da-industria-pesqueira>. Accessed on September 2, 2017.

from fishing.

With a society that is aware of the sector's potential and the benefits it brings, we will be able to reach a new level and seek to break new ground. It is also important to highlight the impact on other areas of society, which, through an even more consolidated economy, can achieve higher quality education and a more robust health system, since as one sector develops, so do all the others.

# 4. METHODOLOGICAL ASPECTS

The research will be carried out using a quantitative and qualitative approach. According to Richardson[12] (1999), quantitative research is characterized by the use of quantification, both in the way information is collected and in the way it is processed using statistical techniques.

According to Trivinos[13] (1987), the qualitative approach works on the data in search of its meaning, based on the perception of the phenomenon within its context. The use of qualitative description seeks to capture not only the appearance of the phenomenon but also its essence, trying to explain its origin, relationships and changes, and trying to intuit its consequences.

The research will be carried out using the descriptive method in which we will observe, record, analyze and order the data, without manipulating it, i.e. without interference from the researchers.

For Cleber Cristiano Prodanov and Ernani Cesar de Freitas[14] descriptive research is:

> When the researcher simply records and describes the facts observed without interfering in them. Aims to describe the characteristics of a given population or phenomenon or to establish relationships between variables. It involves the use of standardized data collection techniques: questionnaires and systematic observation. It generally takes the form of a survey.

It seeks to discover the frequency with which a fact occurs, its nature, characteristics, causes and relationships with other facts. To collect this data, specific techniques were used, such as interviews, using forms and questionnaires.

This study also used the exploratory method, which "aims to provide more information about the subject we are going to investigate, making it possible to define and outline it, i.e. to facilitate the delimitation of the research topic[15] ", by means of a bibliographical survey and a case study.

---

12 RICHARDSON, R. J. **Pesquisa social: métodos e técnicas**. 3. ed. Sâo Paulo: Atlas, 1999.
13 TRIVINOS, Augusto Nibaldo Silva. **Introduction to Social Science Research: qualitative research in education**. Sâo Paulo: Atlas, 1987.
14 Prodanov, Cleber Cristiano. **Metodologia do trabalho cientifico** [electronic resource] : methods and techniques of research and academic work / Cleber Cristiano Prodanov, Ernani Cesar de Freitas. - 2. ed. - Novo Hamburgo: Feevale, 2013, p. 52.
15 Prodanov, Cleber Cristiano. **Metodologia do trabalho cientifico** [electronic resource] : methods and techniques of research and academic work / Cleber Cristiano Prodanov, Ernani Cesar de Freitas. - 2. ed. - Novo Hamburgo: Feevale, 2013, p. 51-52.

Cleber Cristiano Prodanov and Ernani Cesar de Freitas[16] define a bibliographical survey as:

[...] a general overview of the main documents and works carried out on the chosen topic, which other researchers have previously used to obtain data for the research. This bibliography must be able to provide information and contribute to the research. The survey is carried out according to one of two types of research, depending on the topic chosen: the documentary research method and the bibliographic research method.

With the bibliographical survey, it was possible to obtain the necessary data and information for a better analysis of the theme worked on during the research. In this way, the case study, which "is characterized by the in-depth and exhaustive study of one or a few objects, in order to allow a broad and detailed knowledge of them, a task that is practically impossible with the other types of design considered[17] ", also contributed significantly to the better development of the research.

### 4.1 Data collection and analysis techniques

In order to propose a cultural introduction to fishing in the early grades of schools in the Municipality of Itajai and to clarify its importance, interviews were carried out with the former fisheries secretary of the previous administration, the secretary of SINDIPI and a professional who worked for 45 years in fishing.

According to Cervo & Bervian[18] (2002), the interview is one of the main data collection techniques and can be defined as a face-to-face conversation between the researcher and the interviewee, following a method to obtain information on a given subject.

The survey was carried out at the Nilton Kucker State School, with 103 students taking part in the teaching course, where questionnaires were answered with the aim of observing the knowledge of future teachers in the municipal school network on the subject of fishing.

In this way, Manzini[19] (2003) points out that it is possible to plan the collection of information by drawing up a script with questions that meet the intended objectives. The script then served, in addition to collecting basic information, as a means for the researcher to organize

16 Prodanov, Cleber Cristiano. **Metodologia do trabalho cientifico** [electronic resource] : methods and techniques of research and academic work / Cleber Cristiano Prodanov, Ernani Cesar de Freitas. - 2. ed. - Novo Hamburgo: Feevale, 2013, p. 80.
17 GIL, Antonio Carlos. **Métodos e técnicas de pesquisa social**, - 6. ed. - São Paulo: Atlas, 2008, p. 57-58.
18 CERVO, A. L. BERVIAN, P. A. **Metodologia cientifica**. 5.ed. São Paulo: Prentice Hall, 2002.
19 MANZINI, E.J. **Considerações sobre a elaboração de roteiro para entrevista semi-estruturada**. In: MARQUEZINE: M. C.; ALMEIDA, M. A.; OMOTE; S. (Orgs.) Colóquios sobre pesquisa em Educação Especial. Londrina:eduel, 2003. p.11-25.

himself for the process of interaction with the informant.

## 5. THEORETICAL REFERENCE

This chapter deals with important and essential concepts for the development of the project.

### 5.1 Public Fisheries Policies and their Social Implications

Public Policy, as Souza explains[20] (2006), when thought of in managerial terms, is the field of knowledge that needs to put the government into action and/or analyze an action and, when necessary, propose transformations in its course. This normative-institutional and managerial movement takes place under two requirements that qualify the public nature of a policy and its type of appropriation by society: being universal, in the sense that it is valid in its guidelines and materiality for the whole of a nation state, even when it is aimed at a specific public; and acting to guarantee equal conditions and rights for the whole population.

To think about making public policies is to plan for the long term. For these policies to be truly effective, it is necessary to understand the needs of the community and, based on the existing problems, to create social mechanisms that contribute to and alleviate the problems faced by the municipalities in order to guarantee equal conditions and rights for the entire population.

Therefore, in order to produce an effective transformation, they need to stop being just governmental and become real state public policies, which means that they need to be comprehensive, focus on the totality and complexity of society, and not just be restricted to immediate problems, which are much more in the interest of attracting votes than actually becoming something that translates the social interest, thus becoming a political commitment to society.

Há various types of public policies in the fisheries area, such as those of the:

- **Public Fisheries Policies**: those that the government uses to regulate and/or promote fishing activity;
- **Fisheries regulation policies:** aimed at managing the rational exploitation of fisheries resources;
- **Public Policies** to **Promote Fishing:** aim to encourage the development of fishing

20 SOUZA, C. 2006. Public Policies: a review of the literature. In: Sociologias. Porto Alegre, year 8, n° 16, jul/dec 2006, p. 20-45

activity. In Brazil, tax benefits were widely used by the federal government after the 1950s as a policy aimed at regional and sectoral development;

- **Social Public Policies:** aimed at improving the quality of life of artisanal fishing workers, environmental sustainability and food security.

To summarize fishing policies throughout history, with the creation of the Fishing Colonies from 1919 onwards, it was common to find the development of paternalistic policies with medical and dental services as a way of ensuring that artisanal fishermen were linked to the Colony, which, in turn, had been created by the Navy with the aim of guaranteeing a workforce of seafarers for the nascent companies, since its objective was to nationalize fishing and also to guarantee armed arms for possible wars (SILVA, 1988) .[21]

The end of the 1930s saw the establishment of the Estado Novo in Brazil and the intervention of the Getúlio Vargas government in the Brazilian economy, work, culture and social and political life. As Hellebrandt (2012) points out[22] during this last decade, federal initiatives and actions for the fishing sector point to an emphasis on social issues, reflecting both the ideology of the party that took power in 2002 and continues to do so today (PT - Workers' Party), and a worldwide concern with the management of natural resources, as well as food security.

Over the years, fishing has been modernized, increasingly seeking to compete with other sectors of production, but according to Ramalho (2014)[23] , the modernization of the fishing sector has been done without the fishermen's communities, from the top down, in other words, by superimposing on them a class of businessmen who until then had been foreign to local society. In fact, induced modernization, which is extremely favourable in fiscal terms to the fishing industry, is the result of the institutions' view of artisanal fishing, which reflect on it and refer to it from an elitist and capitalist point of view.

Many achievements have been made over time, it is true that a lot has been achieved, but there is still a long way to go in the search for a greater appreciation of the sector and the awareness of society as a whole of the importance of fishing for economic growth, especially in cities where fishing is the main source of income and development, such as the city of

---

21SILVA, Luis Geraldo. **Fishermen in the History of Brazil - Colony and Empire.** Vozes, Recife/PE, 1988.
22 HELLEBRANDT, L. **Conflicts in Artisanal Mullet Fishing in Colony Z3 and their Relationship with Public Policies**. Master's dissertation. Coastal Management. FURG: Rio Grande-RS, 2012, p. 31.
23 RAMALHO, Cristiano Wellington Norberto. **State, Fishermen and National Development: From the Naval Reserve to Aquaculture.** Revista Ruris, Volume 8, n° 1. Campinas-SP, 2014.

Itajai.

## 5.2 Fishing culture

According to data collected, the southern region is the largest producer of fish in Brazil, with the state of Santa Catarina being the largest producer[24] The municipality of Itajai is known as the fishing capital of Brazil. It has a fleet of 500 boats that account for 20% of Brazil's fish production, estimated at 500,000 tons a year. Itajai's fishing production is concentrated on tuna, sardines and shrimp. Around 15,000 people work directly and indirectly in the city's fishing industry[25] .

Public policies for the fishing sector have historically not provided alternatives for the development of the impoverished populations that make up the largest part of the sector.

Fish is a food that belongs to the meat group. It contains a mixture of minerals and vitamins that are essential for a healthy diet. It also has a low fat content compared to other meats, and this fat has a number of benefits in the fight against and prevention of various diseases.

A survey carried out in 2009 by the Ministry of Fisheries found that only 10% of Brazilian municipalities include fish in school meals[26] . For this reason, a project was created in the south of the country to encourage the consumption of fish in school meals, bringing health benefits to students and boosting the fishing economy in the south of the country.

The formation of the city resulted from the contact between two historical and cultural matrices: the fluvial, linked to the river, the land in the interior of the valleys and the riverside populations, and the maritime, associated with the port, the sea and the import/export trade. Both currents have left their mark on the facets of the city's historical development. Fishing and the port contributed to the economic and cultural growth of the local population.

But the waters that brought great positive marks to the town also wreaked havoc in the 80s. Two major floods in 1983/84, on July 8th, started the biggest climatic catastrophe in southern Brazil. The Itajai-Açù River rose 18 meters above its normal level, flooding the entire Itajai valley. The same river that until then had contributed to the city's economy and development left more than a hundred and fifty thousand people homeless and the death toll exceeded

---

24 Available at: http://g1.globo.com/sc/santa-catarina/noticia/2014/03/sc-deve-se-consolidar-como-largest producer of fish in the country.html
25 Available at: <http://itajai.sc.gov.br/c/a-cidade#.Wc-Qu1tSzIU>. Accessed on October 2, 2017.
26 Available at: <http://www.canalrural.com.br/noticias/pecuaria/projeto-sul-pais-incentiva-consumo poixo-merenda-escolar-16458>. Accessed on October 2, 2017.

50. Faced with this situation, the damage left behind was incalculable, and it took the state more than a decade to recover from the tragedy. Before life returned to normal, in a short space of time, less than a year, there was another flood of the same proportions as the previous one.

The history of tragedies is much more recent, and what can be seen is that even after several situations of suffering and catastrophe caused by rivers, nothing has been done to prevent future floods.

# 6. ECONOMY

In terms of the international fishing scene, Brazil's total fish supply ranked 19th in 2010, losing out to other BRICS countries (Russia, 9th; India, 3rd; China, 1st). Brazil's fishing production records include a significant share of extractive fishing, as shown in Graph 2. Extractive fishing, which is made up of artisanal and industrial fishing, has been stable since 2007, even with the recovery in the fishing catch. In 2009, Brazil's extractive fishing production increased its growth deficit in relation to aquaculture by 50%. These results have been increasing over the years.

Graph 2 - Total fish production (t) in Brazil from 1997 to 2011

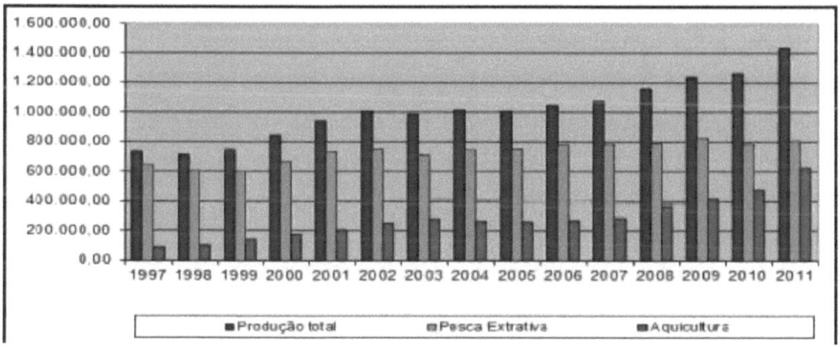


Source: Elabı)raçan pròpria, 2013, cum base πosdadus du IBAMA1 2000-2007; BRASIL, 2008-2011

The increase in consumption of white meat in Brazil can be seen in Graph 4, which shows the upward rise in apparent per capita consumption from 2005 onwards. The demand for meats with a lower fat content has been growing consistently in Brazil and around the world; this moment justifies the Brazilian policy of encouraging aquaculture in favor of extractive fishing. However, even with the encouragement of aquaculture, the fishing trade balance is in deficit, as can be seen in Graph 3.

Graph 3 - Brazil's fishing trade balance, in US$ 1.00 (FOB), from 1997 to 2011

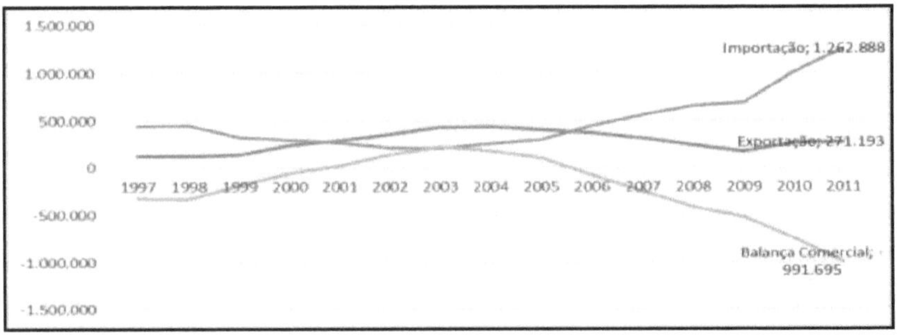


**Source: Own elaboration, 2013, based on IBAMA data, 2000-2007; BRAZIL, 2008-2011**

The deficit in the Brazilian fishing trade balance is recorded from 1997 to 2000.

From 2001 to 2005, the trade balance recorded a surplus, reaching its highest positive balance in 2003 with US$ 224,558. Fish imports exceeded exports in 2006, and this deficit in the fish trade balance widened over the years, reaching US$ - 991,695 in 2011. Even with the encouragement of public policies for fish farming, which have been emphasized since the creation of the Special Secretariat for Aquaculture and Fisheries in 2007, the Ministry of Fisheries and Aquaculture in 2009, and Embrapa Fisheries and Aquaculture in 2009, these efforts have not been enough to meet domestic demand, which has been growing over the years. Apparent per capita consumption in Brazil, according to data from the Ministry of Fisheries and Aquaculture's Statistical Bulletin (BRASIL, 2010), is 9.75 kg/inhabitant/year. However, according to Graph 4, domestic fish consumption is falling while consumption of imported fish is rising. This is due to the fact that Brazil does not catch or produce species with greater economic value on the fishing market, such as cod, salmon, processed fish and others.

Graph 4 - Origin of fish consumed in Brazil from 2000 to 2010

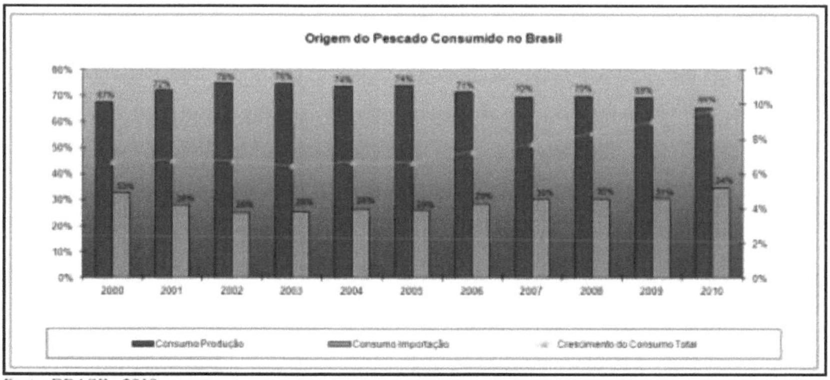


Fonte: BRASIL, 2010

In addition to its share of Brazil's fishing GDP, artisanal fishing generates employment and income for people involved in fishing, making it an important economic activity in Brazil, but one that is not valued by public policies.

The fishing industry provides a significant contribution to Santa Catarina's economy, as well as the continuity, albeit changed, of an important coastal tradition .[27]

---

27 Available at: https://repositorio.ufba.br/ri/bitstream/ri/16344/1/SILVA%2C%20Leidisangela%20Santos%20da.%20A%20ECONOMIA%20PESQUEIRA%20ARTESANAL%20NO%20MUNIC%C3%8DPIO%20DE%20SALVADOR-BA.pdf. Accessed on: 23/10/2017.

## 7. TOURISM

Itajai is located on the northern coast of the state of Santa Catarina, near the mouth of the Itajai-Açu River. The city is located in a strategic region of Santa Catarina, on the right bank of the state's largest river, forming part of an axis connecting it to Brazil's main highways and airports. The municipality has the second largest GDP in the state, due to the most important economic sectors such as logistics, construction, industrial and port activities, with its port complex standing out as the largest fishing port in the country and the second largest in container handling.

The city is constantly growing, especially with new markets expanding, such as the shipbuilding industry and the storage and distribution of fuels. With a population of over 180,000, the city has a considerable historical and cultural heritage. Throughout the year, the city maintains its tradition of typical festivals, which offer culture and entertainment to locals and tourists alike. Itajai also has beautiful natural landscapes, from the countryside to the beaches, crowned by tourist attractions such as Bico do Papagaio, Parque Natural da Atalaia, Molhe, among others.

Fishing in Itajai - Santa Catarina is guaranteed fun, whether on the beaches or in the rivers. There are many options for both amateur and professional fishermen. In the waters of Itajai, sardine fishing is the big attraction in rivers surrounded by spectacular valleys and mountains. Sardines, also known as char, are the cheapest fish you can find. At the Fish Market, a kilo of sardines costs between R$7.00 and R$8.00 and at the Fish Truck R$4.50. The most expensive fish is prawns, which range from R$26.00 to R$90.90, depending on the species[28] All along the 560 km of the Santa Catarina coast there are great spots for spearfishing and ocean fishing, with Itajai being one of the main centers of professional fishing in Brazil. There are also good day and night fishing routes for tourists who visit our region. In the Itajai Valley, there are records of catches within the rules of sport fishing. Also worth noting are the "pesque-pagues" - places where fishing is done in dams or lakes .[29]

28 Available at: <http://itajai.sc.gov.br/noticia/14634/pesquisa-avalia-preco-dos-pescados-em-itajai#.WdFE6VtSzIU>. Accessed on: October 10, 2017.
29 Available at: <http://turismo.sc.gov.br/atividade/pesca/>. Accessed on: 01 Oct. 2017.

## 8. THE IMPORTANCE AND CONTRIBUTION OF FISHING TO THE DEVELOPMENT OF THE LOCAL ECONOMY.

The survey was carried out with some people involved in the fishing sector. The main objective of the questionnaire was to understand the problems faced by the category and what possible improvements could be made by the public authorities and the bodies responsible. The questionnaire was carried out with the president of SINDIPI, a fisherman with 45 years' experience in the fishing industry and an oceanographer who was also Itajai's fisheries secretary in the previous administration.

A questionnaire was also carried out with 103 future teachers from Itajai's municipal education network who are currently attending the teaching course at the Escola de Educaçâo Bàsica Deputado Nilton Kucker.

**Survey conducted with the president of SINDIPI and fishing professionals.**

When asked about the importance of bringing fishing culture to schools, the secretary of SINDIPI emphasizes that it is necessary to present the value that fishing has from an economic and social point of view. The proposal is to get away from the commonplace that has dominated until now, which shows fishing only from an environmental point of view, without considering that it can be practiced very wisely, without affecting the environment.

The oceanographer goes further and says that it is important to spread the word to all levels of society about the importance of the activity, not only economically, but also socially and culturally, and its importance in a special way for traditional communities, quilombolas, Indians and river dwellers, for whom fishing is part of life and tradition.

The fisherman, with his 45 years of experience and no less wisdom, said: "Fishing is certainly an extractive and cultural activity, it is one of the oldest professions in the world, spanning generations, and it is practiced by all the peoples of the earth in various ways, So, since Itajaí is the most important fishing port in Brazil today, and the largest producer of fish from ocean waters, it's only fair that it should be taught as a subject in schools."

The main concern is how much fishing needs to be valued and better structured in the various sectors of society. Faced with this problem, the question that the interviewees answered was: what are the main difficulties faced by fishermen today?

According to the Secretary of SINDIPI, "it's the fishing legislation that treats them as marginals, and in many cases as bandits".

The oceanographer judged the bureaucracy, the lack of recognition of the importance of the activity and the lack of public policies aimed at promoting and developing the sector.

The fisherman observes that today the greatest difficulties faced by fishermen is the poor management of fishing, where government agencies are not renewing the RGPs, licenses and other documentation necessary for fishermen to carry out their work.

When asked about the lack of stimulus from the government. The secretary of SINDIPI considers the lack of investment in the sector to be the main obstacle to its development and maintenance.

The Oceanographer believes that this is reflected in the lack of long-term public policies and the detriment of this productive activity in favor of others.

The fisherman considers that the institutions don't even have the resources to maintain the minimum infrastructure, to keep the processes up to date, so that the fishing sector can function, in addition to the lack of investment in research, lack of availability of resources for investments, costing.

When asked what could be done to improve fishing activity. The secretary of SINDIPI believes that it is necessary to review fishing from the outset, and recompose all the legislation that coordinates it.

The oceanographer says that the activity and its social, economic and cultural relevance need to be recognized, which should be reflected in a state policy and in actions for the continued development and improvement of the national fishing sector, including investments in research and infrastructure to support the activity.

The fisherman, with his professional experience, believes that: Firstly, management must be improved; secondly, investments must be made in minimum infrastructure to keep the sector going; thirdly, the fishing sector must be organized; fourthly, investments must be made in research; fifthly, the rules that restrict fishing activities must be revised.

When asked why fishing has fallen in value over the years. The secretary believes that there is a lack of commitment - in many cases knowledge - on the part of those who should be promoting its sustainable development.

The oceanographer emphasizes the lack of appreciation of the sector, which represents only a small slice of agribusiness in terms of resources generated, and has often been used as a political bargaining chip.

When asked about the devaluation of fishing, the fisherman replied emphatically that yes, the sector does not receive the necessary investment to maintain itself and to provide the necessary support to fishermen.

When considering the importance of fishing for the municipality's economy, the responses were positive with important contributions to the municipality.

The Secretary of SINDIPI believes that fishing brings to the consumer market a product with high added value and nutritional quality, as well as generating many direct and indirect jobs.

The oceanographer provided important data such as: "Every day, fish is gaining more space on consumers' tables as an excellent quality protein. In Brazil, there are more than a million artisanal and industrial fishermen registered with the Secretariat of Aquaculture and Fisheries, who produce around 600,000 tons of fish (data from 2013), and as many as a thousand unregistered fishermen who carry out this activity for their subsistence and that of their families. If we consider the fish production chain, from extraction to the end consumer, and also the family context, which often takes part in the processing of products and the repair of equipment, the activity employs around ten million people. In 2015, the IBGE's Central Business Register recorded that there were 1,788 fishing companies in Brazil (2.63% more than in 2014) and 477 fish preservation and fish product manufacturing companies (4.82% more than in the previous year)".

The fisherman made it clear that it generates hundreds of direct and indirect jobs, generates income, and when fishing is bad, commerce sells much less and the economy in general is felt by the city.

The final question was what is missing for fishing to be recognized as an important economic activity. In addition to emphasizing the importance of unity in the sector, several other suggestions were made.

According to the secretary, the "commitment of the bodies that manage the activity; they are not interested in developing fishing correctly, but in eroding it until it is unfeasible to practice in the country".

According to the oceanographer, in addition to stressing the importance of uniting the productive sector, there must be incentives for the activity, such as credit lines and development programs, the establishment of a sectoral policy with a long-term strategic vision, the valorization of fishery products and sustainable means of production.

The fisherman considered the importance of a union between the fishing sectors and the importance of stimulating the sector's growth in various spheres. He believes that as well as providing jobs, fishing develops various sectors of local society directly and indirectly.

It is understandable that society as a whole is concerned about environmental problems and the extent to which fishing that takes place irregularly and without commitment to the ecosystem can cause incalculable damage to the environment, causing problems that plague the planet and affect the future of societies. The aim is to take a fresh look at fishing and its important social role in terms of fishermen's culture and the economic viability of families that depend on fishing. There needs to be a renewed perspective on fishing activity, to be exercised by all the actors involved, including fishermen, traders, consumers and the whole of society that directly or indirectly ends up benefiting from this sector.

There is a clear consensus that fishing plays a fundamental role in the economic development of Itajai and the region, but it is necessary that the organs

After all, it is unacceptable that even with so much modernization in society, the category is hampered by the problems of poor fisheries management, which make it difficult for fishermen to simply renew their licenses or take a long time to release authorization to fish some species, with fishermen in some cases having to resort to the judicial sphere to authorize the release, making something simple stressful and unnecessary. In this sense, in 2015 the delay in the release of mullet fishing contributed to the high price of mullet and there was uncertainty and low expectations for production in Itajai and the region .[30]

Offering those involved in this activity better working conditions, encouragement and professional development is one of the challenges to be met. The vast majority of fishermen who work directly in the sector do not have the necessary knowledge of legislation and end up being equated with marginals and in some cases, according to the fisheries secretary himself, they are equated with bandits for breaking laws, often out of ignorance and

---

30 Available at: <http://rionegrinhotem.com.br/noticia/8512/pesca-industrial-da-tainha-abre-com-uncertainties-and-low-expectations-for-production-in-itajai-and-region>. Accessed on: October 1, 2017.

prevented from continuing to do what they have done all their lives, which is to fish.

Developing a maritime mentality in our people, consistent with the technological evolution that is currently reaching high levels of development, is a major challenge. Obtaining this constructive mentality is just as fundamental as preserving the sea's resources, because we don't want to stop fishing, but we do want it to be carried out responsibly and sustainably, so that present and future generations can enjoy the whole ecosystem. Bringing fish to families' tables is the ultimate goal of this sector, but in order to have a rational and sustainable exploitation of the riches of our coastal and oceanic zone in the near future, it is necessary, among other things, to design and implement public policies that encompass integrated management planning, balancing and optimizing environmental protection, public use and economic development.

**Future teachers' perceptions of fishing culture in schools**

According to graph 01, when asked if they have ever worked or have someone who works in fishing, the percentage was 67% for those who have not worked or do not know anyone involved in fishing, and 33% for those who said yes. Bearing in mind that Itajaí is one of the largest fishing centers on the Santa Catarina coast, accounting for 59% of fishing production, it is significant that more than half of those interviewed have no involvement with people who work in fishing.

Graph 01

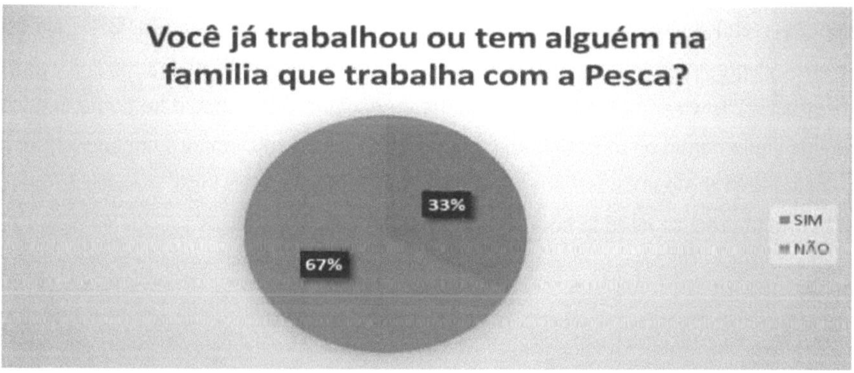


Source: prepared by the authors, 2017.

In relation to the question in graph 02, which asks about the degree of knowledge of the interviewees on the subject of more, 30.29% have moderate knowledge, while 28.27 know little about the subject, 22.21% have minimal knowledge and 12.12% have no knowledge at all. What can be seen is that more than half of the interviewees have knowledge that ranges from little to moderate. This shows the importance of continuing training on the subject, so that planning can be carried out that is committed to the reality of the sector in the municipality, that is efficient and that brings positive results for society.

Graph 02

Source: prepared by the authors, 2017.

Graph 03 shows the possibility of adapting the theme of Itajai's fishing culture to the elementary school curriculum. The percentage of acceptance is 96.93%, almost unanimously agreeing that Itajai's fishing culture should be included in the curriculum, while a very small 7.7% would not implement it in the curriculum. What should be borne in mind is that this implementation must first and foremost be socially relevant, it's not enough for it to be included in the curriculum, there needs to be an objective to fulfill, it needs to bring benefits and the results need to be evaluated and the objectives fulfilled.

Graph 03

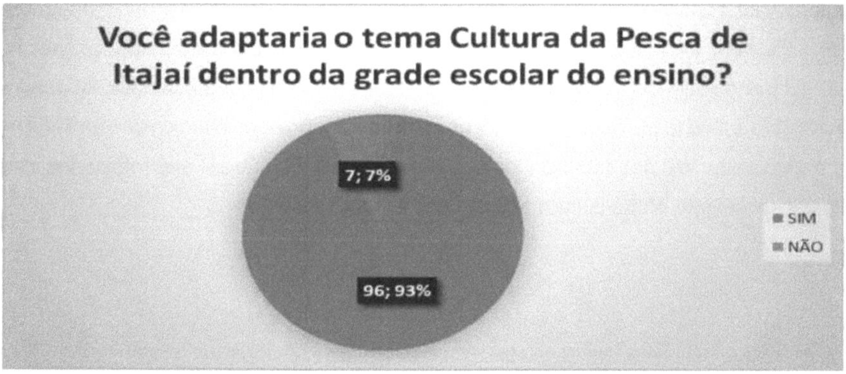


Source: prepared by the authors, 2017.

When analyzing Graph 04, it can be seen that 46.45% of respondents consider culture to be the most relevant topic to work on with students, followed by the environment with 29.28%, and then nutrition

with 28.27%. What is needed is a more general view of fishing and its contributions to the municipality, in the economic context, for example, fishing is of paramount importance to the municipality's growth.

Graph 04

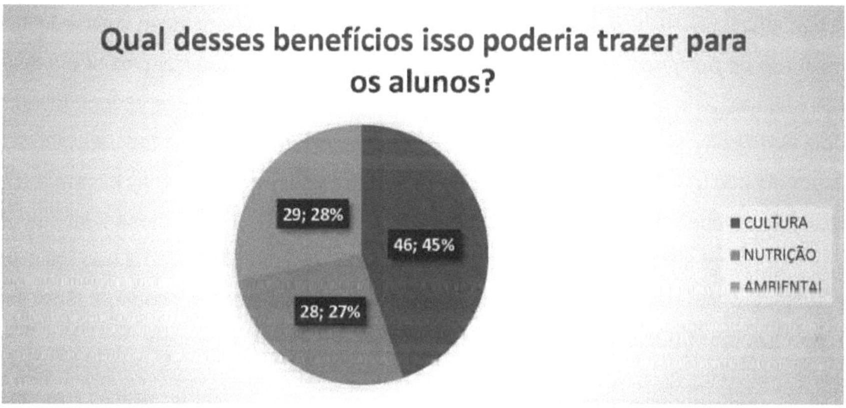


Source: prepared by the authors, 2017.

In relation to graph 05, when asked in which subject fishing could be worked on in the classroom, history was the most cited subject with 38.37% of the answers, followed by geography with 21.20% and arts with 18.17%. Mathematics with 12.12% and Portuguese with 14.14% were also cited. Considering that fishing involves a large number of sectors, both directly and indirectly, it causes a cycle of development to take place and that at least 17 sectors are linked to the fishing industry and it is not only present in the culture and history of the municipality. We need to broaden our knowledge of the subject and rethink the idea that fishing is just part of the culture and discuss it in the classroom.

Graph 05

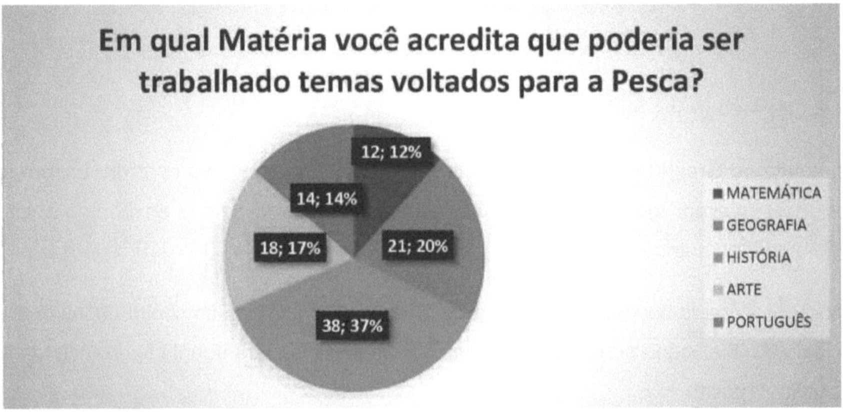


Source: prepared by the authors, 2017.

When something new is proposed, there needs to be commitment from those involved if the results are to be achieved. When thinking about a topic that is not directly part of the lives of those involved, it is necessary to create mechanisms of interest and interaction between the topic and reality so that what is proposed makes sense in practice, in the daily lives of both students and teachers. Work can only be carried out safely and efficiently if there is the necessary knowledge on the subject. According to graph 06, 77.75% of those interviewed have no theoretical knowledge on the subject, while only 26.25% have read a book related to the subject, which requires these professionals to be taken out of their comfort zone; the main objective is to train citizens who understand the sector's role in society.

Seeking to read, understand and appreciate how important fishing is, not just for the municipality of Itajai but for society as a whole, is the duty of every citizen who is concerned with valuing the environment, the economy and development, to understand that one of the oldest professions must be valued and understood as essential so that the product of its work arrives in a nutritious and tasty way and contributes to social growth.

Graph 06

Source: prepared by the authors, 2017.

**8.1 Actions to encourage a fishing culture in the municipality's schools.**

Observing how much the fishing sector needs to be valued by society as a whole, the aim is to propose that the municipality introduce the topic into the curriculum and train teachers from the municipal education network to develop it throughout the school year and in the different subjects, so that the topic is not just remembered on specific dates during the year, but remains under constant debate in classrooms, emphasizing its importance for the economic and social growth of the municipality of Itajai.

**PROJECT:**

**Title:** TRAINING PROGRAM FOR TEACHERS IN ITAJAI'S MUNICIPAL EDUCATION NETWORK

**Target Audience**

The project's target audience includes elementary school teachers from municipal schools in the city of Itajai - SC. These professionals will be responsible for teaching students the importance of cultivating and preserving fisheries in the municipality.

**Justification**

By understanding how important fishing is for the economic, social and cultural growth of the municipality, we realize the importance of preserving this history and also the need to value this profession, which is so old and has contributed so much to local economic growth.

One of the aims of this project is to train teachers in their various areas of activity on the subject of fishing, so that they realize that this subject is extremely important and that it is present directly or indirectly in the city's history.

Professional qualification is essential. Extension courses, lectures and other moments of exposure to subjects such as fishing related to their areas of interest are important for qualification. Libânio[31] (1998) believes that ongoing training leads teachers to reflective action. Pedagogical practice in today's schools requires a teacher who is well trained and prepared to work with students and also with the new problems that are present in society's daily life. As Behrens states[32] (1996, p. 24) "The search for continuing education is necessary for professionals who believe that education is a path to social transformation".

With expanded responsibility, the school today must be able to provide the knowledge necessary for learning, but it must also contribute to the formation of the citizen, so it is necessary to go beyond listing topics for teachers to work on, it is necessary for teachers to be aware of their social responsibility when working in areas that are important and necessary for the municipality.

From this perspective, the role of the teacher - the professional who has direct contact with the student - has been expanded. Their role today is to lead students to an understanding of the importance of what they learn, and to encourage a constant search for knowledge.

Continuing teacher training is another support for teachers to be able to work and carry out their role in society, and to understand how to ensure that their students' time in class is a learning moment.

This project will involve elementary school teachers from public schools, in partnership with the city council and fishing industry organizations. It will be of great importance to the municipality of Itajai-SC, as it will provide a new vision of fishing, and consequently bring benefits to the local population.

---

31LIBÂNEO, José Carlos. **Adeus Professor, Adeus Professora? novas exigências educacionais e profissoes docentes.** Sao Paulo: Cortez, 1998.
32 BEHRENS, Marilda Aparecida. **Continuing teacher training and pedagogical practice**. Curitiba, PR: Champagnat, 1996.

**General Objective**

To design a training program for teachers in the Itajai municipal school system, with the aim of improving and broadening their knowledge on the subject of fishing.

**Specific objectives**

- Identify the most important fishing areas in thc municipality of Itajai and introduce them into the elementary school curriculum;
- To provide teachers with knowledge on the subject, so that they can develop the topic effectively in the classroom;
- To develop ongoing training that will enable participants to understand their role as citizens responsible for raising awareness among municipalities about the importance of fishing for the city.

**Methodology**

**Workload:** 40 hours

**Period**: from February 2018.

| **Operation:** Municipal Department of Education Subjects - MODULE I - content | Workload - 20 hours |
|---|---|
| Regional fishing | 4 hours |
| Regional culture | 4 hours |
| Regional economy | 4 hours |
| Environment and ecology | 4 hours |
| Preserving and raising awareness of fishing | 4 hours |
| Total | 20 hours |
| MODULE II - practical class | Workload - 20 hours |
| Visit to a fish production company | 4 hours |
| A trip on a pirate ship to discover the natural beauty of the region. | 4 hours |

| Visit to the school boat | 4 hours |
|---|---|
| Visit to the oceanographic museum of UNIVALI | 4 hours |
| Seafood tasting | 4 hours |
| Total | 20 hours |

**Teaching staff**

The teaching staff of this project could be a partnership between the University of Vale do Itajai - UNIVALI with the Oceanography course and the IFSC working on the geoscience discipline, involving the disciplines in question.

Table 1 - costs to the municipality

| | Description | Quantity | Unit value | Hours | Total |
|---|---|---|---|---|---|
| | **Coordinators** | **2** | **500 R$** | | **1.000,00** |
| | **Teachers** | | **50.00 R$** | | |
| | **Handouts** | | **35.00 R$** | | **1.750,00** |
| | **Practical lessons** | | | | |
| | **T-shirts** | | **19.00R$** | | **950.0** |
| **Total:** | | | | | **3.700** |

**Source**: prepared by the authors, 2017.

## FINAL CONSIDERATIONS

The proposal for a program to be developed for elementary school teachers by raising awareness of the fishing culture in the municipality of Itajai/SC, presents a compilation of content that contributes significantly to the development of the fishing sector in the region, providing greater knowledge regarding the most different aspects of the industry.

This research sought important information about the importance and contribution of fishing to the development of the local economy, and it was necessary to interact with different professionals in the area in order to understand the problems faced in this sector. Many difficulties are currently faced due to the lack of investment and stimulus in the fishing sector. The aim is to make everyone involved aware of how important and necessary fishing is for the growth of the municipality.

In this way, the aim was to present a new view of the subject, with the aim of introducing it into Itajai's elementary school curriculum, highlighting its importance for local economic development, which has contributed to the city's growth over the years, However, it should be emphasized that fishing is more significant than economic figures. This is the daily life of many people in Itajai, whose tradition has changed, mediated by new technologies and expanded participation, involving various sectors and activities, remaking the fishing culture.

It is worth emphasizing the importance that educators play in this cultural transformation. More than just passing on content, they need to know about the subject in order to be able to convey important and necessary issues in an educational and meaningful way in order to understand the history of fishing and its social contribution to the development of the community, This is why the proposal is to run a training course to work on fishing in schools, trying to ensure that the subject goes beyond the school walls and reaches the homes of the town's citizens, so that they can take a new view of the subject and understand its importance for the whole local community.

It's up to the educator to find ways of including the above-mentioned topic in the curriculum, in line with the syllabus. Forming a link between the contents to be studied, and much more important than the theory of the subject, is the knowledge of how to apply the most diverse situations in a practical and constant way in the day-to-day life of the students, this being a subject that is part of the history of the entire Itajaiense community.

# REFERENCES

PREFEITURA MUNICIPAL DE ITAJAi: Available at: <http://itajai.sc.gov.br/c/a- cidade#.Wc-Qu1tSzlU>. Accessed on October 2, 2017.

PREFEITURA MUNICIPAL DE ITAJAÎ: Available at:

<http://itajai.sc.gov.br/noticia/14634/pesquisa-avalia-preco-dos-pescados-em-itajai#.WdFE6VtSzlU>. Accessed on: October 10, 2017.

Available at: <http://osoldiario.clicrbs.com.br/sc/noticia/2013/01/parque-da-atalaia- em-itajai-e-opcao-para-quem-quer-fugir-do-agito-do-litoral-4006901.html>. Accessed on September 20, 2017.

Available at: <http://rionegrinhotem.com.br/noticia/8512/pesca-industrial-da-tainha- opens-with-uncertainties-and-low-expectations-for-production-in-itajai-and-region>. Accessed on: 01 Oct. 2017.

GOVERNMENT OF SANTA CATARINA: Available at:

<http://turismo.sc.gov.br/atividade/pesca/>. Accessed on: 01 Oct. 2017.

Available at:

<http://turismo.sc.gov.br/institucional/images/phocagallery/costa%20verde%20e%20mar%20-%20itajai/thumbs/phoca_thumb_l_Vista%20A%C3%A9rea%20Molhes%20E%20%20

Praia%20Do%20Atalaia%20-%20Itajai5701.jpg>. Accessed on September 20, 2017.

Available at: <http://turismo.sc.gov.br/institucional/index.php/pt-

br/component/phocagallery/51-costa-verde-e-mar-itajai/detail/359-bico-do-papagaio- itajai-rec0034#>. Accessed on September 20, 2017.

Available at: <http://visiteitajai.com.br/conheca_acidade.php>. Accessed on October 2, 2017.

Available at: <http://www.canalrural.com.br/noticias/pecuaria/projeto-sul-pais-

incentivar-consumo-peixe-merenda-escolar-16458>. Accessed on October 2, 2017.

Available at: <http://www.litoraldesantacatarina.com/itajai/caracteristicas-de-itajai.php>. Accessed on October 2, 2017.

Available at: <http://www.mgmoperadora.com.br/destino/2285/1/28/destinos-nacionalitajai>. Accessed on October 2, 2017.

Available at: <https://ndonline.com.br/florianopolis/noticias/producao-registrada-between-2010-and-2013-keep-santa-catarina-in-the-leadership-of-the-fishing-industry>. Accessed on September 2, 2017.

Available at: <https://ndonline.com.br/joinville/noticias/calculo-do-pib-catarinense-e-revisado-pelo-ibge-e-coloca-joinville-itajai-e-florianopolis-no-topo>. Accessed on September 21, 2017.

HELLEBRANDT, L. **Conflicts in Artisanal Mullet Fishing in Colony Z3 and their Relationship with Public Policies**. Master's dissertation. Coastal Management. FURG: Rio Grande-RS, 2012.

MANZINI, E.J. **Considerações sobre a elaboração de roteiro para entrevista semi-estruturada**. In: MARQUEZINE: M. C.; ALMEIDA, M. A.; OMOTE; S. (Orgs.) Colóquios sobre pesquisa em Educaçâo Especial. Londrina:eduel, 2003.

MOREIRA, Jani Alves da Silva, and LARA, Angela Mara de Barros. **Public policies for early childhood education in Brazil** (1990-2001) [online]. Maringà: Eduem, 2012. 246 p. ISBN 978-85-7628-585-4. Available from SciELO Books.

PIMENTA, CAM., and ALVES, CP., eds. Politicas pùblicas & desenvolvimento regional [online]. Campina Grande: EDUEPB, 2010. 211 p. ISBN 978-85-7879-016-5.

Prodanov, Cleber Cristiano. Metodologia do trabalho cientifico [electronic resource]: methods and techniques of research and academic work / Cleber Cristiano Prodanov, Ernani Cesar de Freitas. - 2. ed. - Novo Hamburgo: Feevale, 2013.

RAMALHO, Cristiano Wellington Norberto. **State, Fishermen and National Development: From the Naval Reserve to Aquaculture.** Revista Ruris, Volume 8, n° 1. Campinas- SP, 2014.

SILVA, Luis Geraldo. **Fishermen in the History of Brazil - Colony and Empire.** Vozes, Recife/PE, 1988.

SOUZA, C. 2006. **Public Policies**: a review of the literature. In: Sociologias. Porto Alegre, year 8, n° 16, jul/dec 2006.

# APPENDIX A - QUESTIONNAIRE USED TO COLLECT DATA UNIVERSIDADE DO VALE DO ITAJAÎ - UNIVALI

## PROESDE COURSE

The questionnaire is being used to complete the PROESDE course work.

Interview with future teachers

Age:

( ) 18 to 25 ( ) 26 to 31( ) 32 to 40( ) 41 to 49( ) Over 50 years old

Gender:

( ) Female ( ) Male

1 - Have you ever worked or does anyone in your family work in fishing?

( ) YES ( ) NO

2 - How much do you know about fishing culture?

( ) None ( ) Minimal ( ) Little ( ) Moderate ( ) Quite a lot

( ) Excellent ( ) Mastery

3 - Would you adapt the theme of Itajai's fishing culture within the elementary school curriculum?

( ) YES ( ) NO

Because:

4 - What benefits could this bring to the students?

( ) Cultural

( ) Nutritional

( ) Environmental

( ) Other

Which ones?

5 - Did the school you go to or attended during your schooling mention any topics about Itajai's fishing culture and its importance?

( ) YES ( ) NO

6 - In which subject do you think we could work on fishing issues?

( ) Mathematics

( ) Geography

( ) History

( ) Art

( ) Portuguese

7 - Have you read any books on fishing culture?

( ) YES ( ) NO

## Interview with fishing representatives in Itajai

1. How do you see the fishing culture in Itajai today?
2. Do you think fishing is important for the town's economy?
3. What are the main difficulties faced by fishermen today?
4. Is there a lack of investment by the government in the fishing sector?
5. What could be done to improve fishing activity?
6. Do you think that bringing fishing culture to schools contributes to their development?

## ANNEX A - DECLARATION OF AUTHENTICITY

### DECLARATION

I hereby declare, for all legal purposes, that I take full responsibility for the ideological contribution given to this work, being aware of the provisions of Law No. 9610 of 18/02/1998, exempting the University of Vale do Itajai - UNIVALI, the coordination of PROESDE DEVELOPMENT and the content advisor from any and all responsibility for it.

of 2017.

Student's name: ______________________________

Signature: ______________________________

### ANNEX C - Signature of those responsible

SIGNATURE OF THOSE RESPONSIBLE

Name of academic

GUILHERME ANTONIO FABRICIO CARDOSO

LARISSA NUNES

LUISA BARBIER SANTOS

ROZANGELA APARECIDA DE MELO GONÇALVES

Content supervisor Prof.ª Msc Josiane da Luz Prof. Msc Ayrton Santos

Course leader

Prof. MSc Samara Aparecida da Silva Garcia

MIX
Papier aus verantwortungsvollen Quellen
Paper from responsible sources
FSC® C105338

Printed by Books on Demand GmbH, Norderstedt / Germany